Mattawan Elementary
School Library

GIANTS of Smaller Worlds
drawn in their natural sizes

Birdwing butterfly - Troides mirandus

Unicorn beetles - Dynastes tityus

Camel Cricket - a species of Schizodactylus

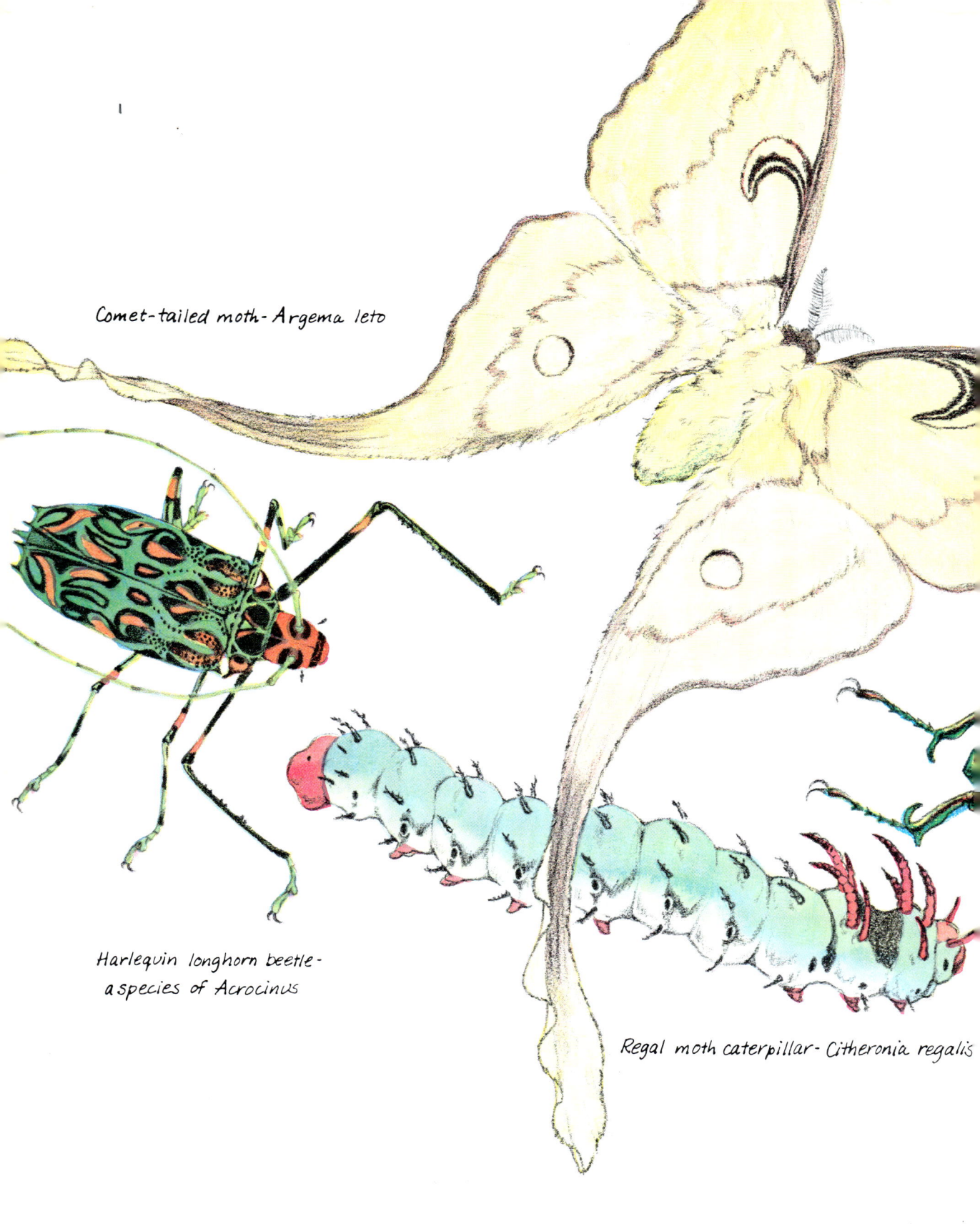

Mattawan Elementary
School Library

GIANTS of Smaller Worlds
drawn in their natural sizes

Joyce Audy dos Santos

Dodd, Mead & Company   New York

Scarab beetle — Chrysophora Chrysochloa

Hissing cockroaches — Gromphadorhina chopardi

Longhorn beetle — Psalidognathus frendi

*To my uncle, Roger Houle,
because his arms are always open wide*

Copyright © 1983 by Joyce Audy dos Santos
All rights reserved
No part of this book may be reproduced in any form
without permission in writing from the publisher
Distributed in Canada by
McClelland and Stewart Limited, Toronto
Manufactured in the United States of America

1 2 3 4 5 6 7 8 9 10

Library of Congress Cataloging in Publication Data

Dos Santos, Joyce Audy.
   Giants of smaller worlds.

   Includes index.
   Summary: Introduces some "giant" members of the
arthropod family, including ponerine ants from Brazil,
stag beetles from Europe, the bird wing butterfly of
New Guinea, and the giant scolopender from South
America.
   1. Arthropoda—Juvenile literature. [1. Arthropods]
I. Title.
QL434.D67   1983   595.2   82-45993
ISBN 0-396-08143-5

Atlas moth - Atlas attacus

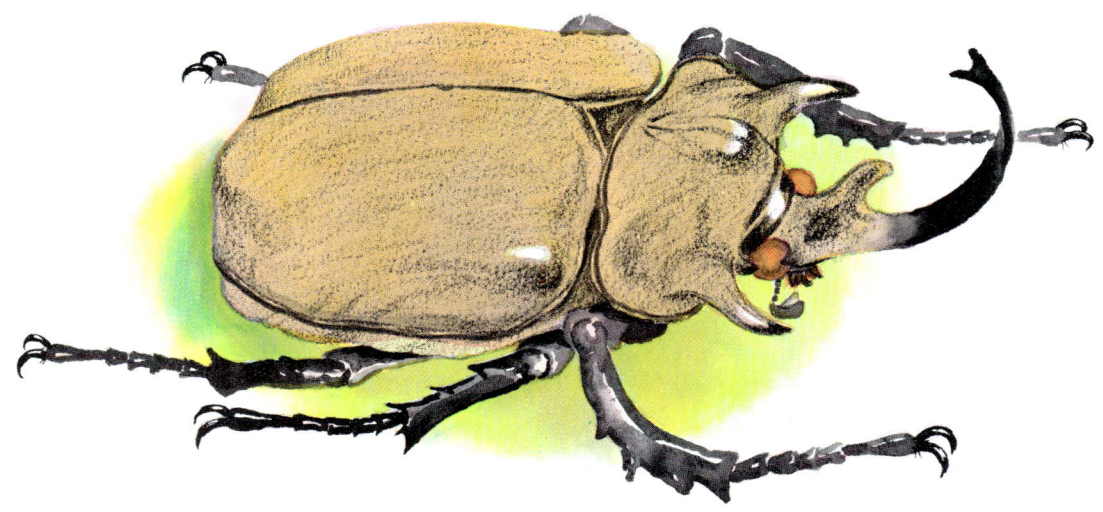

Elephant beetle - Megasoma elephas   Top view

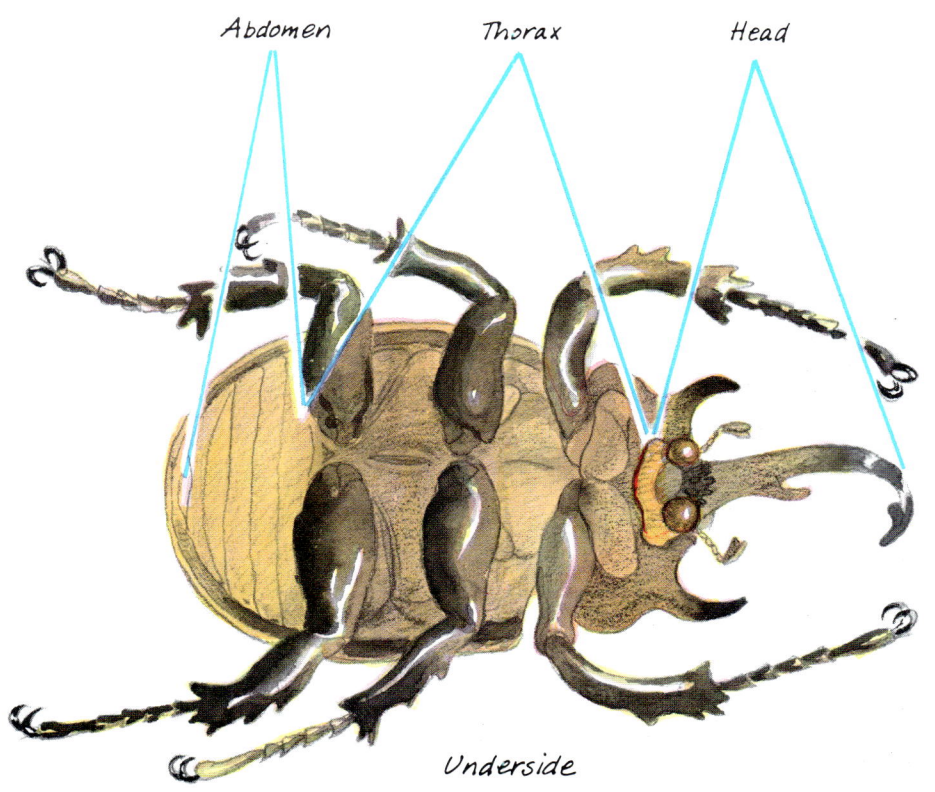

Abdomen   Thorax   Head

Underside

A special group of animals lives all around us. They look and act differently, but they are all alike in certain important ways.

All of them have legs with connecting joints. All of them have bodies that are divided into separate parts. And all of them have skeletons that are, unlike ours, on the outside of their bodies. These hard outer skeletons, called exoskeletons ("exo" means "outside"), help protect the animals.

Insects, spiders, scorpions, centipedes, millipedes, crabs, and lobsters are all members of this group. They are called arthropods, a name that means "jointed feet."

Insects are arthropods that have six legs and three body parts: a head, a thorax, and an abdomen. They are the only arthropods with wings. Spiders all have eight legs and two body parts. Scorpions also have eight legs but more than two body sections. Centipedes and millipedes have many legs and many body parts.

This book is about some of the biggest insects and their arthropod relatives.

All of the arthropods in this book are drawn in their natural sizes. Put your hand down next to any picture. How does it compare? Could you hold the real arthropod in your hand?

The arthropod you see here is a Goliath beetle. Goliath beetles are insects. In fact, they are the heaviest insects in the world. Some weigh nearly a quarter of a pound. They look scary, but they are harmless. They feed on sap and nectar from banana and palm trees.

When Goliath beetles fly, they are noisy. In parts of Africa, children will tie a Goliath beetle to one end of a string and knot the other end to the tip of a stick. When the beetle tries to fly away, it becomes a strange buzzing toy—until it is finally let go.

Any animal reaching for this beetle had better beware. The Goliath beetle can use its body like a trap. It lowers its head and thorax, leaving a space between the thorax and the hard wing covers. When the Goliath beetle snaps back, anything caught in that space—like a monkey's finger—will be pinched.

Goliath beetle - Goliathus giganteus

The insects on this page are all ants.
These tiny ones are leaf-cutting ants.

Leafcutting ants - Atta texana

These carpenter ants are more than half an inch long.

Black carpenter ants - Camponotus pennsylvanicus

These big ones are ponerine ants from Brazil. They hunt and eat insects and sometimes other ponerines. The ponerine ants are sixteen times longer than the small leaf-cutting ants.

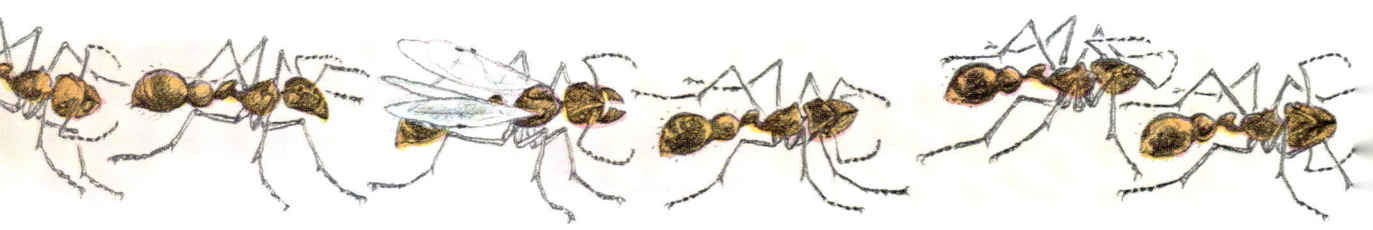

Ponerine ants - Paraponera clavata

What would happen if some children were sixteen times taller than others? One child would be four feet tall and another would be sixty-four feet tall, more than three times higher than a giraffe, which is the tallest land animal in the world.

Actually, no one could really be sixty-four feet tall and still have a body designed like ours. If we were sixteen times taller we would be much more than sixteen times heavier. Bones like ours would be crushed by all that weight. A heart like ours couldn't pump hard enough to circulate blood that far up.

Whether insects are big or small, they all need certain things in order to live and grow.

They need the right climate. These giant cockroaches live in tropical Central and South America, where even the winters are sunny. They grow well in their warm environment. In places where winters are cold, most insects live in special ways to survive the harsh weather.

Giant cockroaches - *Blaberus giganteus*

Stag beetles - Lucanus Taiwanus

All insects also need a good supply of the food that is right for them.

At night these stag beetles climb and fly among oak and chestnut trees in search of sap. Their huge horns are mouth parts, but they are not used for chewing. They have become pincers, which are used at mating time. When one male stag beetle meets another, each uses its strong pincers to try to toss the other out of the tree. The beetles don't often get hurt, because of their tough exoskeletons.

Stag beetles like these are the largest beetles in Europe.

This giant water bug hunts tadpoles, salamanders, and small fish in the shallow parts of streams and ponds. It spears them with its beak, but it does not eat by chewing, as beetles do. Instead, it softens its victim's body by injecting digestive juice. Then it sucks the nutritious liquid meal down.

The giant water bug cannot breathe underwater. It breathes by backing up to the surface and sticking the two breathing tubes at the end of its body into the air. Its back legs are shaped like paddles to help it swim.

Male giant water bugs carry the female's eggs piggyback. The female cements them on his back, and he swims with them until they hatch.

Giant water bugs can be found in South America and Asia. Smaller ones live in this country.

Beware. They are also called toe-biters.

Giant water bug - Lethocerus maximus

Queen Alexandra birdwing - Ornithoptera alexandrae

High in the treetops of New Guinea, flowering plants stretch toward the sun. Fluttering about their blossoms, huge birdwing butterflies feed on nectar. One species, the Queen Alexandra birdwing, is the largest butterfly in the world.

The warm sun and sweet nectar help the birdwings reach their incredible size. But there is something else besides the right food and the right climate that all living things need for growth, and that is water.

What keeps the soft body of the birdwing caterpillar firm and round? Water does. The caterpillar is inflated by the water in its body.

Insect blood is mostly water. When the birdwing butterfly emerges from the cocoon it spun as a caterpillar, its wings are folded, wet, and tender. The butterfly pumps blood through the veins on its wings to stretch them to full size.

If you gave an average size butterfly plenty of warmth, food, and water, would it grow as big as a Queen Alexandra birdwing butterfly? It would get bigger than other butterflies of its kind that did not have ideal conditions for growth, but it could never grow to the size of the birdwing. The parents of the birdwing were very large. And their parents were large. And their parents before them.

Birdwing caterpillar and emerging male

Queen Alexandra birdwing—*Ornithoptera alexandrae*

It is not known exactly how and when each type of giant insect got to be so big. Some kinds may have been big since they first appeared on earth. Others may have started out small. But if the larger ones in a population were better at getting food and defending themselves than the smaller ones, then they would be more likely to survive and bear offspring that were also large. Some of the giants in this book may have evolved to their large sizes over many generations.

Mammoth cicada—Pomponia adusta

A large insect is not automatically a better survivor than a small insect. Most insects alive today evolved to sizes that are right for the different ways they live, eat, and defend themselves.

More than fifteen hundred kinds of cicadas of different sizes, colors, and life-styles have evolved. Each type fits well into its own small world. Some cicadas found in the lush jungles of Malaysia are enormous. Other tropical cicadas are smaller than the periodical cicadas of the eastern United States.

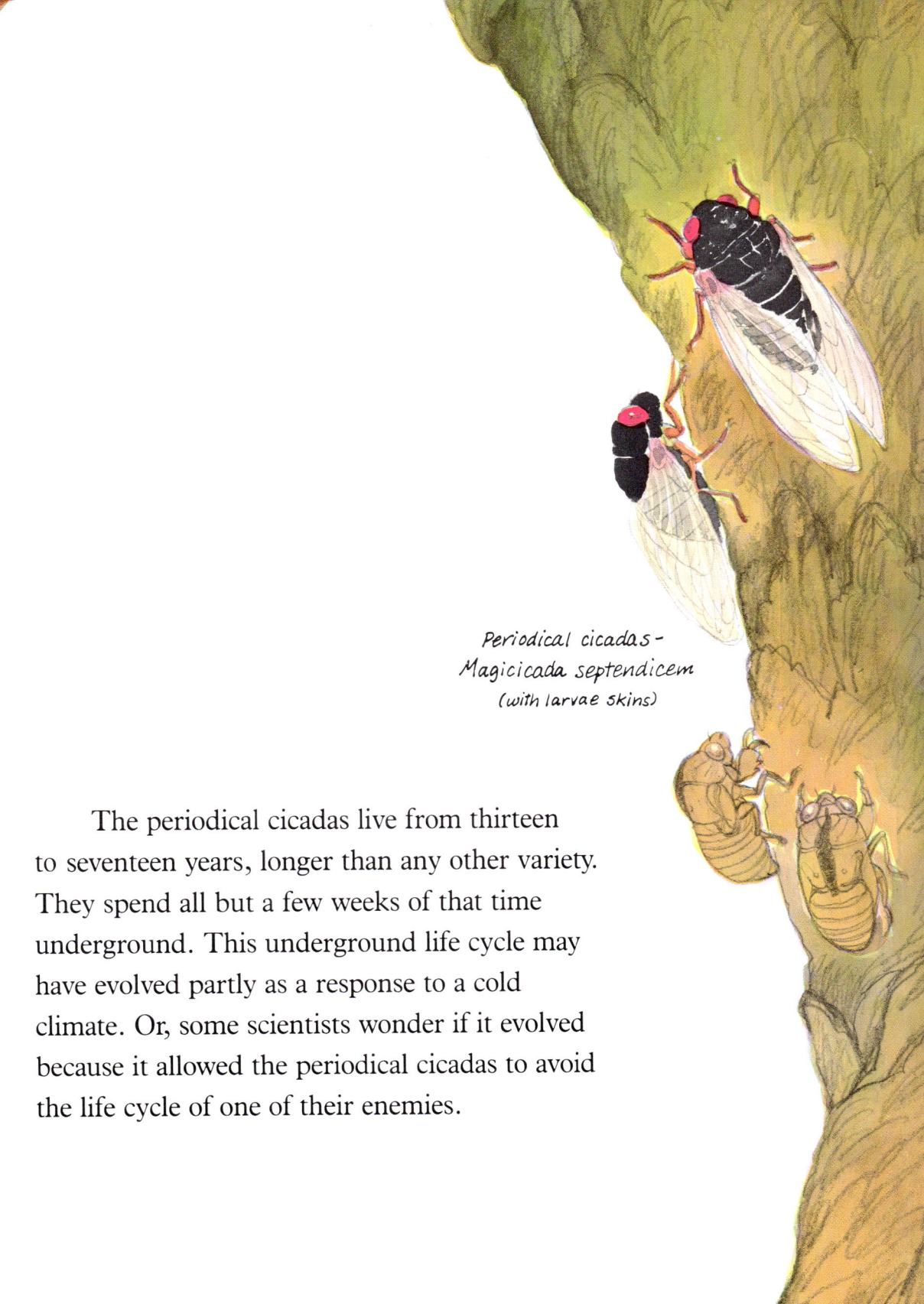

Periodical cicadas —
Magicicada septendicem
(with larvae skins)

The periodical cicadas live from thirteen to seventeen years, longer than any other variety. They spend all but a few weeks of that time underground. This underground life cycle may have evolved partly as a response to a cold climate. Or, some scientists wonder if it evolved because it allowed the periodical cicadas to avoid the life cycle of one of their enemies.

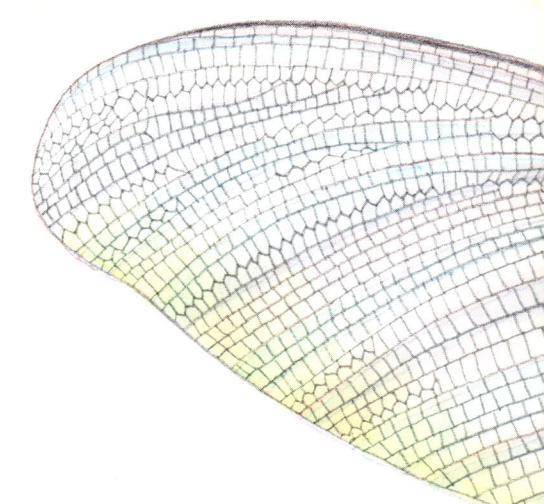

Dragonflies had a big surprise in their evolution. This one is a shady aeshnid. It hunts insects near shady woodland streams. Like other dragonflies, it has huge compound eyes that help it see better than many other insects. When a shady aeshnid sees an insect it can eat, it will hold its legs down like a basket. Then it swoops in and scoops up its prey.

Millions of years ago, before there were dinosaurs, relatives of dragonflies were zooming over ponds and swamps. Those prehistoric dragonfly-like insects were excellent hunters, too. Many were about the size of dragonflies today. But in their lush, warm environment some evolved to an astonishing size. Turn the page to see how large they got to be.

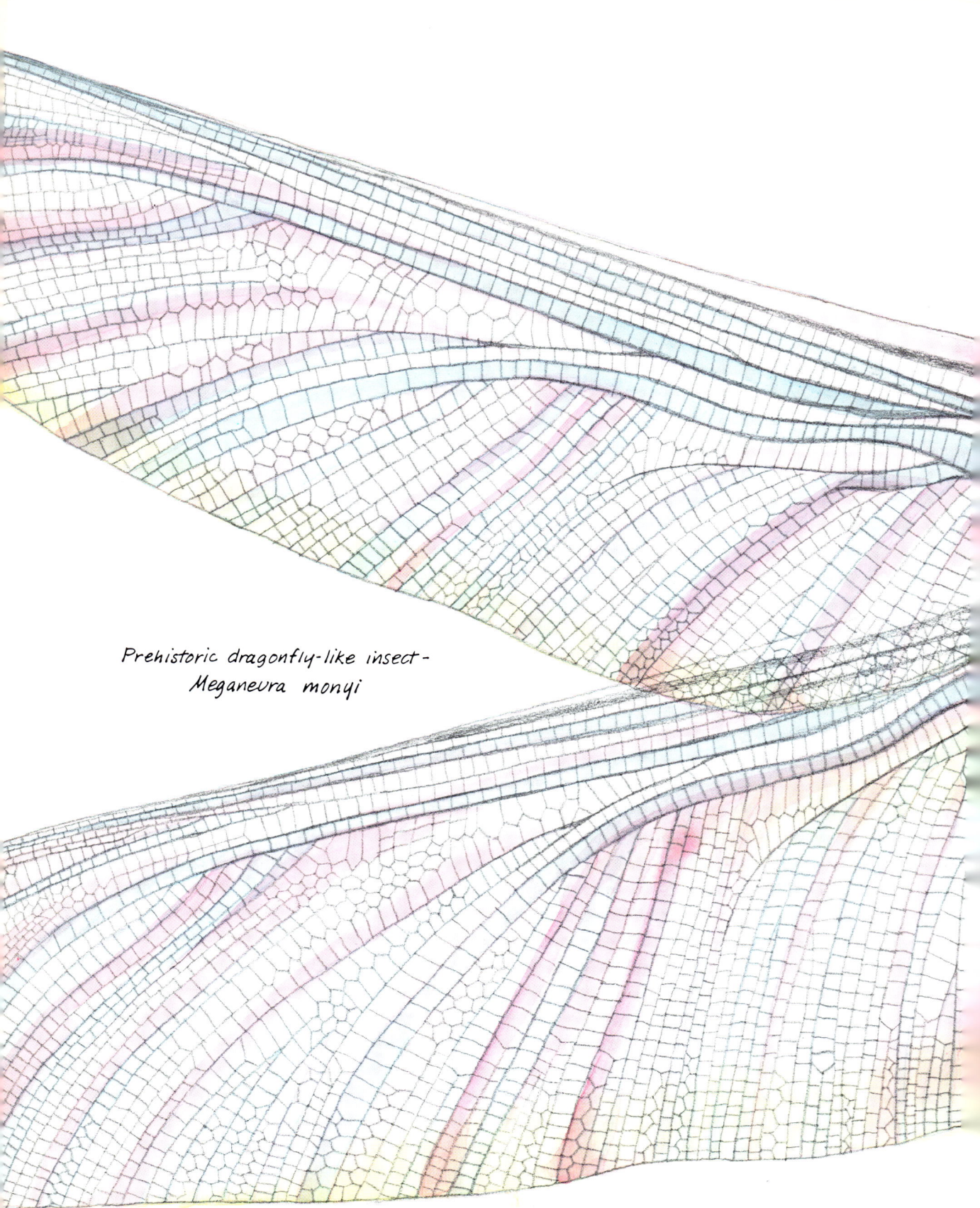

Prehistoric dragonfly-like insect - Meganeura monyi

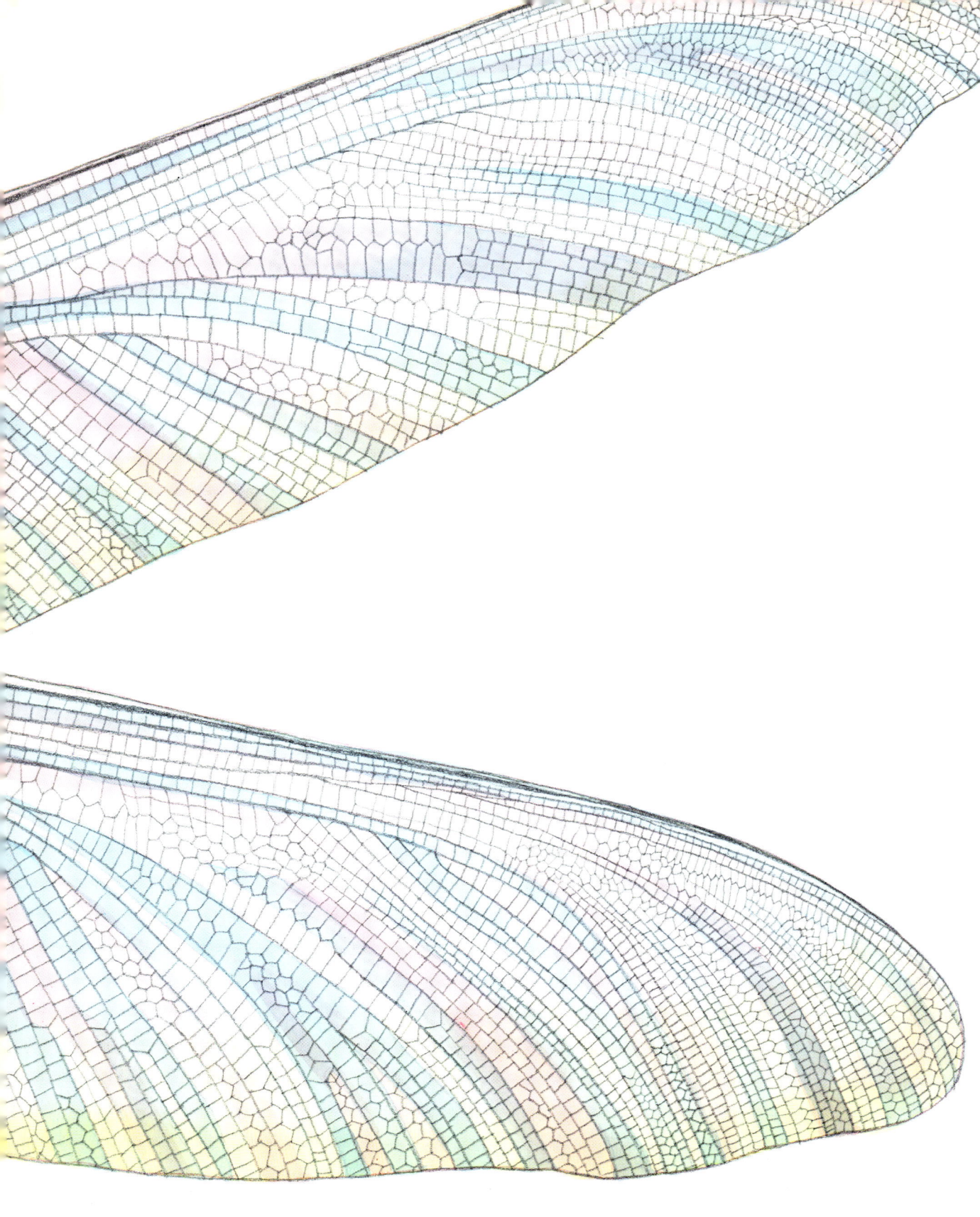

Why aren't there dragonflies with twenty-nine-inch wingspans today? No one knows for sure. Did something happen to the food they ate? Some people think that other flying animals appeared on earth and ate them. Another part of the mystery is that no fossils of the young have been found. Perhaps their bodies did not preserve well. Or there may have been animals that ate the young, causing the giant dragonfly-like insects to become extinct.

Prehistoric dragonfly-like insects were good fliers. But not all big insects are. These elephant and Hercules beetles don't always dart away when they are in danger, as many smaller insects would. But the elephant and Hercules beetles have tough exoskeletons to help protect them.

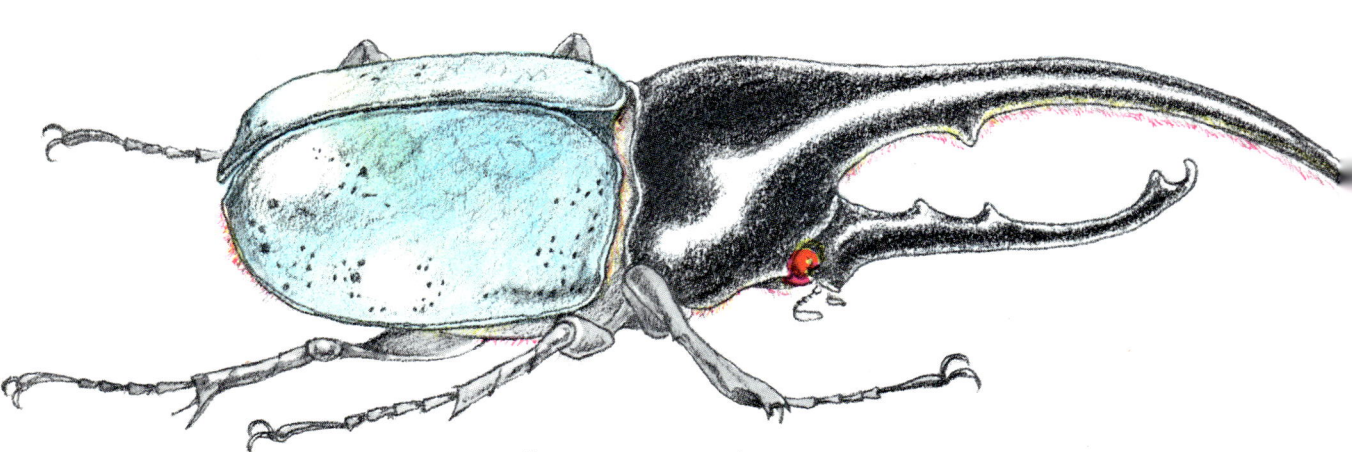

Hercules beetle - Dynastes hercules

Elephant beetles - *Megasoma elephas*

The horns on these beetles are not mouth parts, as they are on the stag beetles. The bottom horn grows out of the beetle's head, and the top horn grows out of the thorax. When the beetles move their heads, the horns open and close like pincers. The males use them in battle at mating time. The beetles' mouths are very small and hidden under their heads.

What else besides their tough exoskeletons do big insects use for protection? It is not always easy for a large insect to find a place to hide. Many of them don't move quickly, either. How do they escape being eaten?

Walkingsticks like this one grow to be the longest insects in the world. Even though some have bodies thirteen inches long, they are experts at hide and seek.

Walkingstick - Diagoras ephialtes

At night when many insect-eating animals are asleep, walkingsticks climb through the trees feeding on leaves. During the day they seem to disappear among the twigs and branches. Their crusty brown or green exoskeletons and twiggy legs blend in perfectly.

Walkingsticks are protected by their camouflage. They really do look like sticks walking—except when they stand still.

Other giant insects use camouflage, too. They often look like the plants they feed on. There are five insects hidden here. Can you find them?

Some large insects use more than camouflage for protection. A lantern fly will stay very still and blend in with its surroundings. But if it is threatened, then—flash—the lantern fly spreads its wings and a pair of scary eyes appears. Suddenly the lantern fly looks like part of a bigger animal.

Lantern fly—a species of Laternaria

The tip of this Atlas moth's wing resembles a snake's head to fool the moth's enemies. Its colors also act as camouflage when the moth perches among the leaves that collect at the base of a tree. Like the wings of all moths and butterflies, the Atlas moth's wings are thin, clear membranes covered with colored scales. But they also have triangular "windows" where there are no scales. The color of the background shows through, making the Atlas moth hard to see.

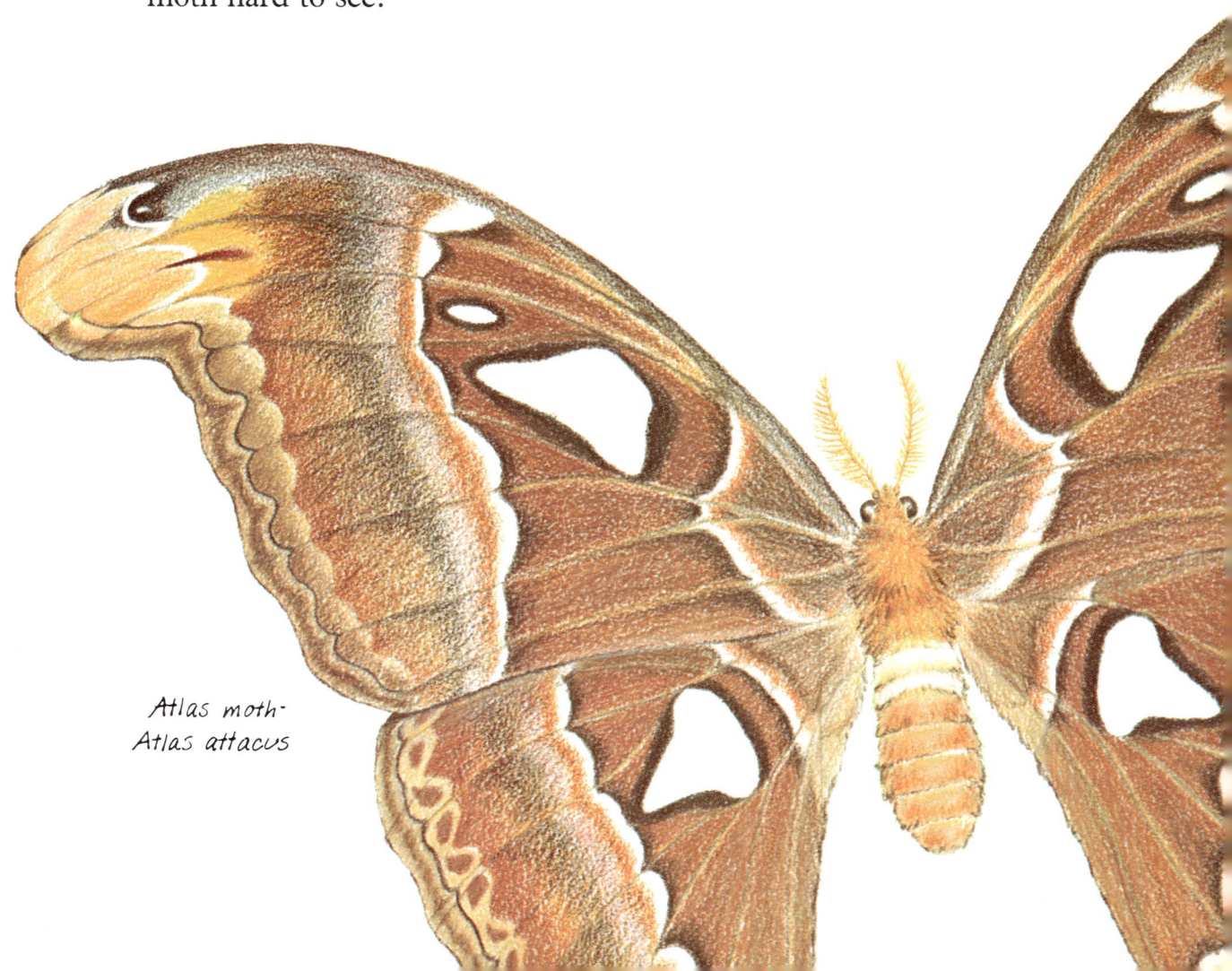

Atlas moth-
Atlas attacus

Giant scolopender - Scolopendra subopinipera

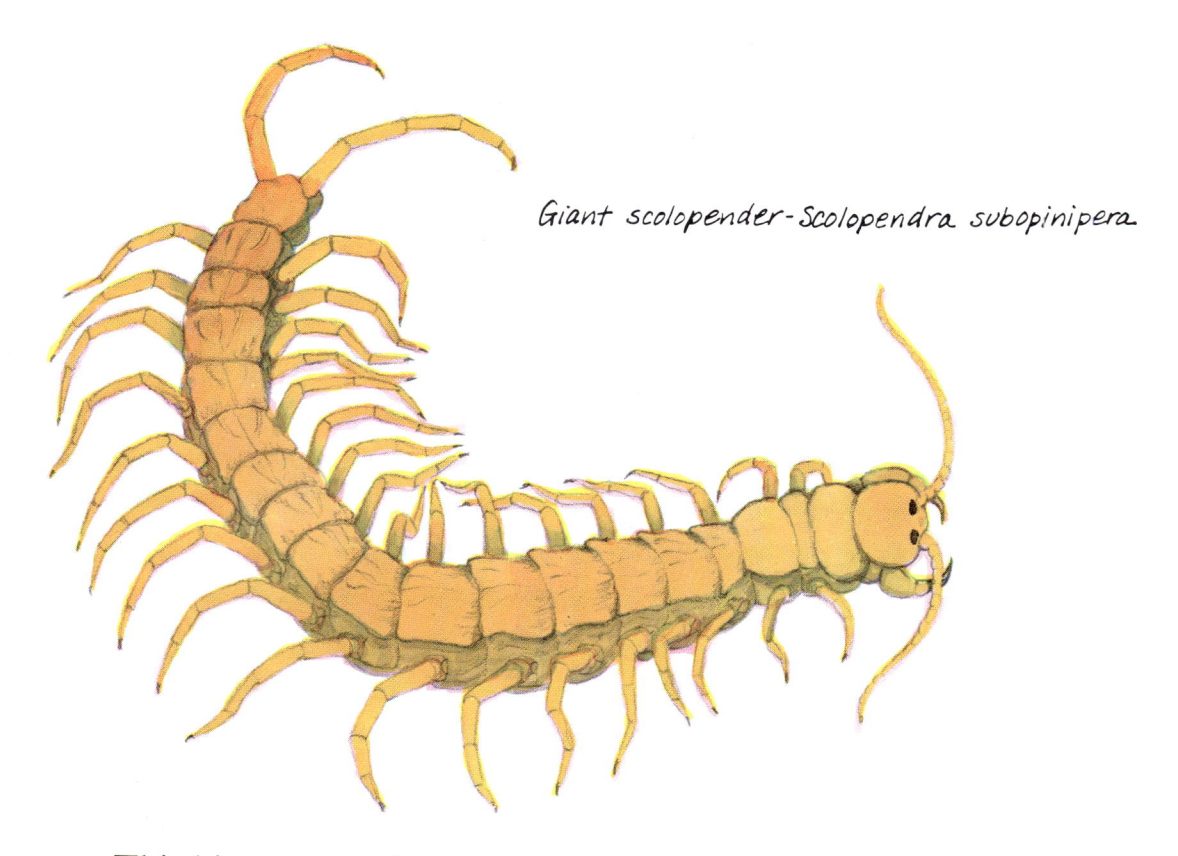

    This big arthropod does not need camouflage for protection. It is a centipede from South America called a giant scolopender. It uses its first pair of legs like poison fangs. They are hollow, and poison flows through them.

    The giant scolopender hunts cockroaches, lizards, and other small animals among the wet, decaying leaves on the forest floor. It must be a fast runner to catch its food.

    Centipedes are not insects. They have lots more than six legs and three body parts. Every centipede has one pair of legs on each section of its body except the first and last sections. Some centipedes have as many as 173 pairs of legs.

Giant millipedes like this one have two pairs of legs on most body sections. Millipedes also live on the forest floor, but they eat plants, not animals. They protect themselves by emitting a poisonous liquid with an awful smell. An insect trapped in a container with a millipede can actually die from this bad-smelling substance.

Giant millipede - a species of myriapoda

This arthropod is also not an insect. It is a bird-eating spider from South America. It feeds on insects, lizards, snakes, small mammals, and little birds.

Bird-eating spider- a species of Avicularia

No one believed the first person who reported seeing a spider eating a bird. In 1705, Maria Merian published a book of her nature paintings. One of them showed a large spider dragging a hummingbird from its nest. Everyone who saw the picture thought it was impossible for spiders to behave this way. But 158 years later, another naturalist saw spiders killing finches. Then people knew that Maria Merian had been telling the truth.

The bird-eating spider stabs its prey with its fangs and injects venom, which helps to soften the body. The spider adds digestive juice, and when the victim's body is almost liquid, the spider drinks it in. The bite of this spider is not deadly to people, although some of the hairs on its body can seriously irritate a person's skin.

Scorpions use poisonous stings when they hunt and to defend themselves. This huge scorpion lives in rocky areas in Africa, where people sometimes die of scorpion stings. It looks like it would be dangerous, but in fact its sting isn't deadly. The sculptured *Centruroides* scorpions of Mexico and the American Southwest have been known to kill people, even though some of these scorpions are only one-quarter inch long.

Scorpions hunt insects and spiders at night. During the day they hide under rocks or logs or in burrows they have dug. In places where scorpions are common, people must shake out their shoes and clothing before putting them on.

Scorpion— Pandinus imperator

Rhinoceros beetle- Dynastes centaurus

Why don't insects and their relatives get even bigger than they are? Could a rhinoceros beetle ever grow as big as a rhinoceros?

Bodies of arthropods are very simple. The air the rhinoceros beetle breathes passes into its body through tiny tubes, called tracheae, that open at many places on its exoskeleton. The tracheae carry the air the short distance to the tissues. If the beetle were much bigger, it would need a system of many tubes to reach farther and farther into the tissues. It would also need something to speed up the flow of oxygen.

If the rhinoceros beetle were very much larger, its exoskeleton would no longer give it enough support. It would have to develop some sort of better support on the inside. It would also need a much stronger pump to circulate its blood farther.

foot of a black rhinoceros

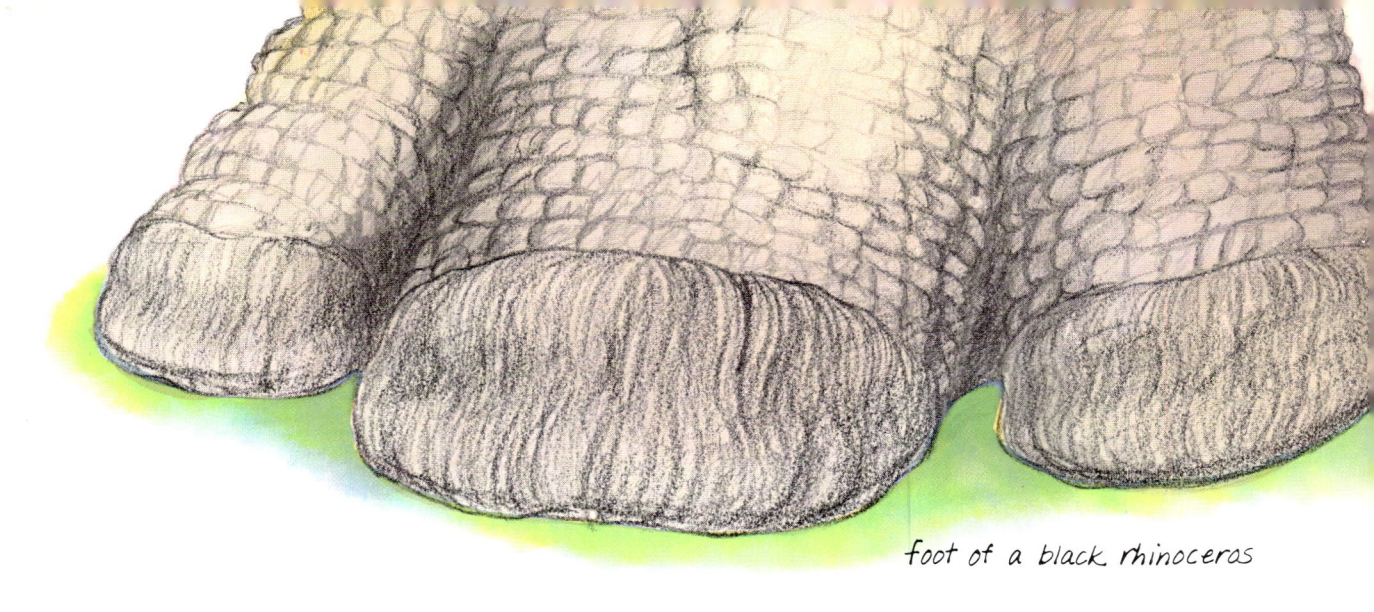

    Crabs and lobsters can reach larger sizes partly because they live in water. The largest arthropods alive today are Japanese spider crabs. Some measure twelve feet from tip to tip of their slender, outstretched claws, although they are only fifteen inches across their bodies. Because they live in the sea, their weight is supported by the water around them. Land arthropods like insects would have to change very much to survive at larger sizes. In fact, to survive at a larger size, an arthropod would have to change so much that it would not be an arthropod anymore.

    A scientist named J. B. S. Haldane once wrote an essay called *On Being the Right Size*. "The higher animals are not larger... because they are more complicated," he said. "They are more complicated because they are larger."

    Arthropods may be simpler and smaller than many other animals, but some are giants in their own small worlds.

Backyard Superbugs

Many large insects live in tropical areas. But others can be found almost anywhere. How good a hunter are you? These are some you can find in the United States.

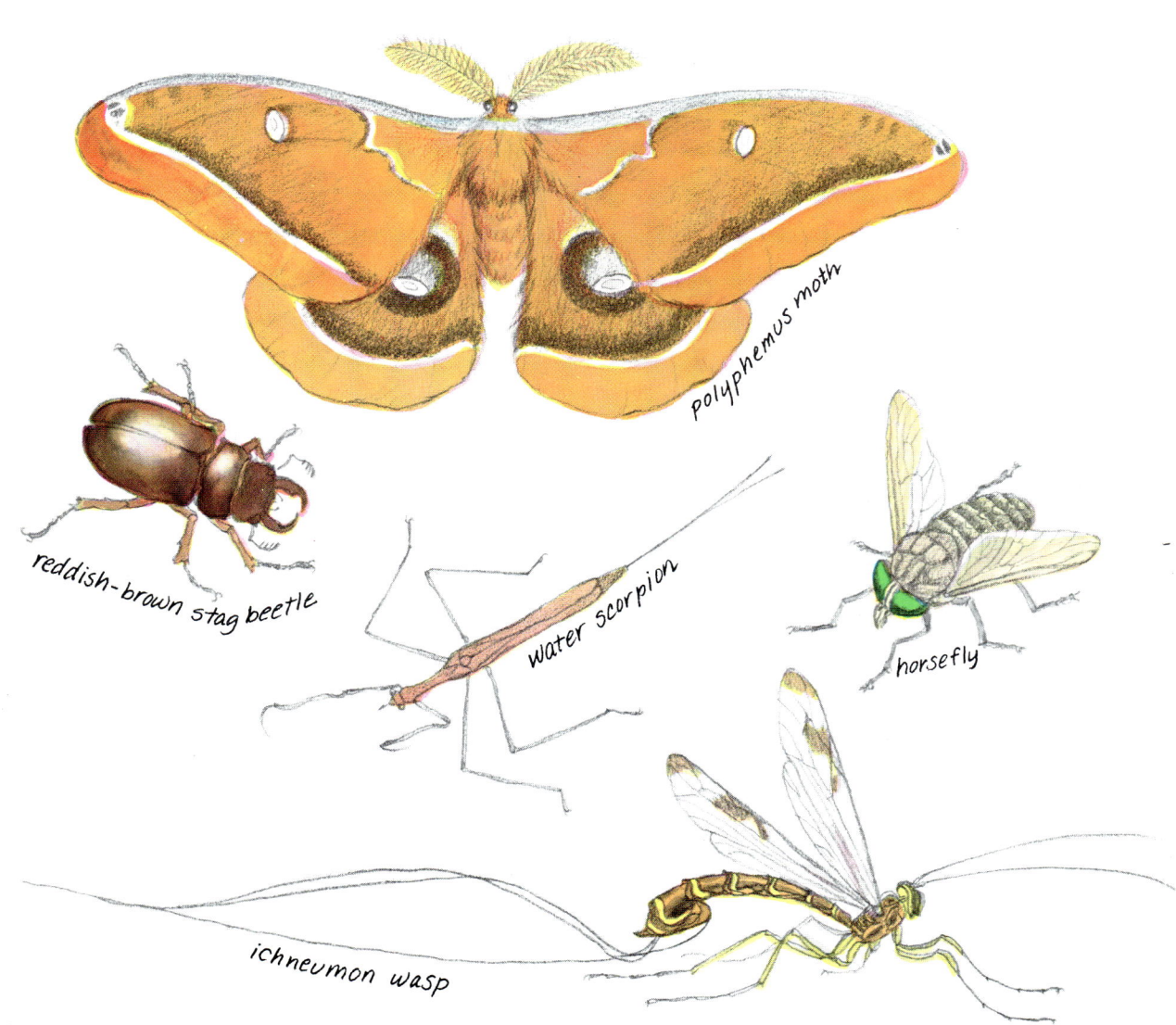

polyphemus moth

reddish-brown stag beetle

water scorpion

horsefly

ichneumon wasp

45

Acknowledgments

There are many people to whom I owe my gratitude for extending their time and advice during the evolution of this book.

At the Museum of Comparative Zoology at Harvard University, those who found specimens for me to draw were Dr. Al Newton and his assistant Margo Grimes in the Entomology Department; Dr. Herbert W. Levi and his assistant John Hunter, specializing in spiders in the Invertebrates Department; and Dr. Frank M. Carpenter in the Fossil Insect Department. Dr. Carpenter offered not only his warm hospitality and very helpful visual material, but was also kind enough to read the manuscript. The text was also read by Lou Sorkin at the Museum of Natural History in New York. I am indebted as well to Paul Miliotis, a Research Associate who specializes in dragonflies at the Museum of Comparative Zoology.

At the Boston Museum of Science, Don Salvatore, the Assistant Curator of Collections, allowed me access to two very intriguing specimens, and Dick Sheffield of the Exhibits Department loaned me visual reference material.

Bill Rowe of the Entomology Department at the Smithsonian Museum of Natural History in Washington, D.C. assembled some incredible insects, and the live Insect Zoo at the Museum proved inspiring as well. Two of the insects included in the book were drawn from living residents of the zoo.

May I also express sincere thanks to friends, many of them children, who brought me specimens or lent me books. Among them are Linda Bourke, Nancy Rutherford, Linda Gilmartin, Aaron, Tara, Scott, Rubén, and my children, Carl, Eric, and Melody.

The insects camouflaged on pages 32-33 are: top left, Macleay's spectre (Extatosoma tiaratum, *female); bottom left, metallic woodboring beetle* (Catoxantha opulenta); *bottom center, Malaysian bush katydid; bottom right, leaf insect* (Phyllium siccifolium); *top right, longhorn beetle* (Macrodontia cervicornis).

Grey witch moth - Thysania agrippina

47

Index

Aeshnid, shady, 23
Africa, 8, 40
Ants, carpenter, 10; leaf-cutting, 10; ponerine, 10
Arthropods, 7-8, 36, 38, 42-43
Asia, 14
Atlas moth, 5, 35

Beetles, elephant, 6, 28-29; Goliath, 8-9; grapevine, 45; harlequin longhorn, 2; Hercules, 28-29; large diving, 45; metallic woodboring, 32; reddish-brown stag, 44; rhinoceros, 42; rose, 43; stag, 13, 29; unicorn, 1
Bird-eating spiders, 38-39
Birdwing butterflies, 1, 16-19
Brazil, 10
Bush katydid, 32
Butterflies, 35; birdwing, 1, 16-18; birdwing, Queen Alexandra, 16-19; tiger swallowtail, 45

Camel cricket, 1
Camouflage, 31, 32, 34-35
Carpenter ants, 10
Caterpillars, birdwing butterfly, 17; regal moth, 2
Centipedes 7, 36; giant scolopender, 36
Central America, 12
Centruroides scorpion, 40
Cicadas, 20-21; mammoth, 20; periodical, 20-21
Cockroaches, giant, 21; hissing, 3
Comet-tailed moth, 3
Crabs, 7, 43; spider, Japanese, 43
Cricket, camel, 1

Defenses, 8, 13, 21, 28, 29, 30-35, 36, 37, 40
Dragonflies, 22-28; prehistoric, 23-27; shady aeshnid, 22-23

Elephant beetles, 6, 28-29
Europe, 13
Evolution, 19-27
Exoskeleton, 7, 28, 30, 31, 42

Fossils, 27

Giant cockroaches, 12
Giant millipedes, 37
Giant scolopender, 39
Giant water bugs, 14-15
Goliath beetles, 8-9
Grapevine beetle, 45
Grey witch moth, 47
Growth, 10, 17, 19, 20, 23, 42, 43

Haldane, J. B. S., 43
Harlequin longhorn beetle, 2
Hissing cockroaches, 3
Horsefly, 44

Ichneumon wasp, 44
Insects, 7, 8, 10, 12, 13, 20, 28, 32, 36, 38

Lantern fly, 34
Large diving beetle, 45
Leaf insect, 33
Leaf-cutting ants, 10
Lobsters, 7, 43
Longhorn beetle, 3, 33

Macleay's spectre, 32
Malaysia, 20
Mammoth cicadas, 20
Merian, Maria, 39

Metallic woodboring beetle, 32
Mexico, 40
Millipedes, 7, 37; giant millipede, 37
Moths, 35; Atlas, 5, 35; comet-tailed, 2; grey witch, 47; polyphemus, 44

New Guinea, 17

Periodical cicadas, 20-21
Polyphemus moth, 44
Ponerine ants, 10
Prehistoric dragonfly-like insects, 23-27

Queen Alexandra birdwing, 16-19

Reddish-brown stag beetle, 44
Rhinoceros beetle, 42
Rose beetle, 45

Scarab beetle, 3
Scolopender, giant, 36
Scorpions, 7, 40-41
Shady aeshnid, 22-23
South America, 12, 14, 36, 38
Spider crab, 43
Spiders, 7, 38; bird-eating, 38-39
Stag beetles, 13, 29

Tiger swallowtail, 45

Unicorn beetle, 1
United States, 20, 44

Walkingsticks, 30-31
Water bugs, giant, 14-15
Water scorpion, 44
Wolf spider skin, 45